WHEN TELEVISION IS A MEMBER OF THE FAMILY

WHEN TELEVISION IS A MEMBER OF THE FAMILY

Edward N. McNulty

ABBEY PRESS
St. Meinrad, Indiana 47577

PHOTO CREDITS: Cover, E. McNulty; p.16, Jean-Claude Lejeune; p.37, © Copyright 1977. Tandem Productions, Inc. All Rights Reserved; p.62, Family Communications, Inc.; p.78, Wm. Koechling.

Library of Congress Catalog Card Number
79-55860
ISBN: 0-87029-160-2

©1981 St. Meinrad Archabbey
St. Meinrad, Indiana 47577

CONTENTS

Preface	7
1. The Challenge of Television	13
2. TV's Impact on the Family	20
3. Role Models on Television	30
4. TV and Your Family's Values	41
5. Children's Programming	56
6. The TV "Advangelist"	67
7. Coping with Television	73
Suggested Resources	89

Preface

Television is a member of the average family in the industrial world. It is both a loyal servant and, at times, a hypnotic tyrant. It entertains in a manner more lavish than that enjoyed by the emperors of Rome. It propagandizes more people than Dr. Goebbels ever could. And it challenges us, sometimes in a forceful way, to be concerned and involved in the world's problems.

There are many books written about television. Some of these are about the programs and the stars, others about the sociological and psychological impact of TV, and others analyze it as a medium of communication. This isn't just one more book *about* television. The "Member of the Family" is an important part of the title. We will be thinking about television from the standpoint of family dynamics—how does the medium influence the family, and how can families cope with it?

We will be examining and discussing television from the perspective of what it means to be a servant of Jesus Christ in the electronic age. Long before the CBS eye lit up our screens, the Apostle Paul wrote, "Whatever you do, do all to the glory of God." That covers a lot of territory—and in the last quarter of the twentieth century, even our televiewing.

We'll be examining television's role in the family from the point of view of the Christian gospel with its emphasis on life-affirming values and its concerns for human relationships. At times we'll have to look at some threats posed by television, issue some warnings, and discuss some strategies for dealing with the dangers. However, this is not an anti-TV book. We'll also focus on the positive aspects of television—its potential for supporting and enhancing family relationships and its possibilities for enlarging our experience and sense of oneness with the world. Some of our finest hours as a family have been spent gathered in front of the TV set, watching and then discussing a drama or situation comedy.

As members of local churches we have an additional reason to be concerned about television: our responsibility for the young in those communities. In most Christian communities that practice the baptism of infants, certain promises are made. The parents affirm their faith and promise to live and teach the faith so that the child will one day confirm what was done for him or her at baptism. Because we are a faith *community,* parents do not bear this responsibility alone. In some churches the entire congregation, and in others godparents who represent the People of God, make a promise to look after the welfare of the child. Since children are the most ardent and vulnerable consumers of television, our Christian responsibility must include concern for children and television. What values are they picking up from their ten to thirty hours of televiewing a week? What are they learning about what it means to be a man or woman from their favorite TV heroes or friends? How can parents and other adults help children get *more* out of *less* TV? These, and many more, are some of the issues with which this book deals.

This book is written from the experience of one who has matured along with television and is now a parent of five teenagers. I can remember the last of the "golden days" of radio and the excitement we all felt when the first television station came to town. As a lover of opera as well as rock, I've never been able to understand the cultural snobs who sniff discernably whenever the subject of television enters a conversation and proudly proclaim that they never watch the "boob tube." Granted that there is much garbage in video land, I have always found much to challenge my mind and spirit, as well as just to provide a few hours of relaxed escape from a pressure-cooker schedule.

As parents, my wife and I have had to deal with the question of what to allow our children to watch. We've been concerned over the years with the effects of TV advertising on our children's values, especially when the message of the TV adman was in conflict with what we were attempting to teach and live. We've sometimes had to look at our own viewing habits as parents. An example of this was our watching the evening national newscasts, particularly in the sixties and early seventies. The 6:00 news was our time for catching up with the exciting events of the world and vicariously being a part of them (made all the more necessary from our standpoint because we were living in a small isolated community). However, the news program came on during the dinner hour. This meant either that we hurried through dinner or that we watched while we ate. Either way our children were being shortchanged. We were rushing through the meal and cutting short family conversations. Something had to give. For us it was television. We might not be as well informed about national and international events, but the relaxed sharing time during dinner is worth it as we take the opportunity to become informed of our sons' and daughters' interests and activities.

Television can provide some really great moments for families. I'll always be grateful to the producers of "The Autobiography of Miss Jane Pitman" and "Roots." These provided our children with valuable insights into the background of our country's racial problem. These programs gave us an opportunity to talk with our children about prejudice and racism in a nonpreachy way. I was able to share with my children some of my experiences in civil rights work. We were also able to talk about the prejudices of some of their friends and classmates and to discuss ways of dealing with them. We were able to do this easily and naturally because we had shared a common TV experience.

I hope that parents, pastors, educators, and those concerned with family relationships will find in this book some helpful insights and techniques. Television and the family is a subject that only recently has attracted the interest of the churches, except for a few voices crying in the wilderness years ago. The majority of priests, ministers, and religious educators still preach and teach as if television had never been invented (or in too many cases, as if they *wish* that it hadn't!). When TV has been noted, it has often been in a self-defeating negative manner. The case in the summer of 1978 with the series "Soap" is a good example. Fearing its exploitation of promiscuous and perverted sex, some religious groups played into the hands of the show's producers by mounting a noisy campaign to force the network to cancel the show. This only increased the interest of the general public, so much so that the series became a hit. "Soap" might very well have gone the way of most mediocre new series if it had not attracted so much publicity.

There is a lesson to be learned in this, as parents attempt to regulate TV. I suspect that those parents who ban TV from the house are making it all the more attractive to their children who are bound to see it while visiting friends' homes. Instead, if television is treated as a cherished yet somewhat dangerous member of the family, parents will be able to deal constructively with it. May this book make such coping both fun and profitable.

CHAPTER ONE

The Challenge of Television

Have you ever arrived at the checkout counter of your supermarket and discovered that your six-year-old daughter has made her own deposits in the grocery cart—a box of Krinkly Krunchies, two packages of Yummy Candy Bars, a can of cherry Fresh Aide, and three packets of Bubbl-O-Gum? Or did you ever have to pull your screaming child away from the huge discount store display of the latest stunt toys as he sobbed, "But, Mother, I want one! I *need* it!" Do you arrange your schedule around certain TV shows, and find yourself checking *TV Guide* before phoning a friend at night? After all, you don't want to commit the serious social error of interrupting a favorite program! Are you or your kids wearing a T-shirt, scarf, hat, gym shoes, or who-knows-what bearing the name of a commercial product or a TV/Hollywood star's visage? If the answer to any of the above is "yes," then you're very much caught up in our media-saturated culture.

But you're not alone for no one can escape the effects of our electronic media, and of television in particular. Even those who pride themselves on not owning a television set are affected — they just don't realize it. Before 1950, television was in only a few thousand homes. In 1950, 4,000,000 homes welcomed their first television set. By 1955, 67% of American homes had a set; by 1960, 88%; by 1965, 92%; and today almost every household has a TV.

So taken for granted is the ubiquitous TV set that statisticians are now counting the households with two, three, or more sets. Perhaps you or your neighbor are giving little twelve-year-old Debbie Dimples her very own television set. Now she can watch her favorite programs without hassling other members of the family. Debbie's reaction? Disappointment. She's crestfallen because it's a small black and white rather than a large screen color set. "How can you expect me to watch 'Laverne and Shirley' and the movies in black and white?" she exclaims through quivering lips. Poor deprived child. Her dream solution: have the family buy one of the new giant screen models (only the price of a new car a few years back) and let her have the family's "old" color set.

Little Debbie — and millions of "Johnnies" too — watch TV for entertainment. And so do most of their parents. But TV can do more than provide diversion from the schedules and harassments of daily living. At its best, TV can be that window on the world long claimed by the leaders of the TV industry. Great events of history and sports are brought to us. Great plays and adaptations of novels. The feeling of what it was to be a slave. The aching sense of growing old and knowing that death is not far away. The problem of a mother and daughter communicating across the chasm of hostile years. These experiences and much more, are conveyed to an audience of millions by the so-called "boob tube."

Although we had begun to realize the power of television during the national elections of the 1950s and of 1960, it took the death of a young president to make us aware how potent the medium could be. When the news from Dallas flashed across the land, most people turned on their TV sets not just for the latest word, but for the latest picture. In an earlier age we would have rushed out for newspaper extras. In a tribal society we would have gathered around the village drummer to await the news. Our TV sets in a sense made us a tribe again—but it did more than merely bring news. It allowed us to see the young widow and child, the swearing in of a new president, and the solemn dignity of the funeral that attracted the heads of state from all over the world. TV calmed us, reassuring us that the nation's affairs would be carried on in firm, new hands.

TV also assured us that we would survive other traumas and revolutions of the 1960s. Indeed, one of those revolutions, the civil rights movement, TV helped to foster and spread. Most white Americans, north or south, infected with

racism had long ignored the plight of their fellow black Americans. TV would not let them do so any longer, once Martin Luther King, Jr., successfully led the bus boycott in Montgomery, and college students began their series of sit-ins. Such dramatic confrontations drew TV cameras like honey attracts bears, and there on our screens we were confronted with the ugly face of our racism. Millions watched Dr. King's great speech during the march on Washington; and none could escape the almost nightly reports of the beatings and fire-hosing of men, women, and children; the bombings of schools, homes, and churches; and the jeering, taunting crowds confronting little children at school. We didn't like the results of racism, so Congress submitted to our pressure to pass new laws.

This certainly did not bring about the dream of "liberty and justice for all." Some progress in equal rights and the related war on poverty became lost in an ambivalent reaction to the Vietnam war and anti-war activism in the 70s, yet the 70s were not the 50s. No television station would dare put on "Amos 'n Andy." "The Jeffersons" and "Good Times" presented a far healthier view of Blacks than did films and TV shows of an earlier age. And most powerful of all was the message of the two series of "Roots." The extreme winter and the fuel shortage of 1977 shut down many of the nation's schools, so that many families watched the episodes together. "Roots" meant much to black people in terms of instilling pride and dignity into children for the endurance and heroism of their ancestors—and it meant much to white viewers as well. Just after "Roots II" was shown, a white businessman expressed his appreciation for the series and concluded, "During the 60s I didn't understand what all the protests were about. Now I understand why Blacks were so mad—and still are."

Most Americans have never seen a live production of a Shakespearian drama, yet millions will be able to see the entire plays of Shakespear, thanks to the massive British effort to tape them. Even though shown on the less than popular Public Broadcasting System, as many people will witness the Bard's plays as have seen them live since they were first written!

Children can now move clear across the country and find that they have much in common with the youngsters in their new neighborhood, for all have been watching the same comedies, cartoons, or adventure shows on TV. We'll leave to later any judgment as to whether it's good or not that they are so familiar with "Charlie's Angels," "CHiPs," or "The Incredible Hulk." Nevertheless, they will know these wherever they live. Television, in one sense, is the great unifier of our time. Electronically, we are being joined together as one tribe sharing the same myths and legends.

Some critics have even suggested that televiewing is now the basic religion of our culture. It has become a ritual for millions to watch it everyday starting with "The Today Show" or "Good Morning, America," picking up after work with the prophet figure, Walter Cronkite, continuing through the evening with a potpourri of situation comedies and adventure shows, and concluding the day with the high priest of the bedroom, Johnny Carson, or on one of his many "off" nights, a late show.

Others refer to such televiewing as a habit—a drug habit. They argue that the viewer is addicted to television, unable to live comfortably without it, often experiencing withdrawal symptoms when unable to get to a television set. According to such critics, television is indeed a dangerous drug—a "plug-in drug." We'll look at television from these two perspectives in the following chapters, noting at times the shortcomings of such views, and at other times, recognizing the helpfulness of looking at TV from such perspectives.

CHAPTER TWO
TV's Impact on the Family

Anything that consumes a large amount of your family's time needs to be examined and talked about, whether it be a family vacation, chores around the house, the employment of family members, or recreation activities. So, take a few minutes, gather the family together, and make a family viewing profile.

In a column on the left of a piece of paper, write down the hours of the day. Across the top write the days of the week. Then in the appropriate spaces, write the names of the TV programs that your family watches. If the TV set is on while the family is engaged in other activities, indicate this also. Have a copy of *TV Guide* handy so that you can look up the name of a program you may not be able to recall. You might be surprised that you can't recall details of many of the programs, either the names or the story lines. If so, that should tell you something about the quality of your TV experience!

To provide a basis for comparison, write in the appropriate time slot your other activities — work, school, hobby time, shopping, personal grooming, listening to records, volunteer service in your community, church, reading, etc. Total up your TV viewing hours, then your other activities. How do the totals compare? Are you shocked? How many hours are you devoting to your growth in mind, body, and spirit? How much of your time benefits others as well as yourself? What opportunities do you have for interaction with other people and for the exchange of ideas.

These are challenging questions, and they can lead to others when, as family, you discuss your TV viewing habits. The drawing of a family viewing profile could cause a re-evaluation of your personal approach, and that of your family, to the role of TV in your home. If you and your family are watching less than 10 or 12 hours a week, you probably don't need to read the rest of this book. But if you're the "typical American family," the set is on at least 4 to 8 hours a day.

In many homes, especially where both parents work at outside jobs, television is used to keep the kids occupied after school. Questions about such a practice should be raised, concerning both the amount of televiewing by the children and the quality of what is watched. The argument can be made that irregardless of the quality of shows watched, the excessive amount of televiewing that some children engage in can be harmful.

In 1977, this story was wired to newspapers across the country:

> MIAMI—The defense of a 15-year-old boy charged with killing his elderly neighbor was based upon "involuntary television intoxication," his lawyer says.
>
> In a pretrial brief filed yesterday, lawyer Ellis Rubin said Ronald Zamora "was suffering from and acted under the influence of prolonged, intense, involuntary, subliminal television intoxication" at the time.
>
> "Through the excessive and long-continued use of this intoxicant, a mental condition of insanity was produced," Rubin's brief argued, rendering Zamora unable to realize "the criminality of his conduct. . . ."
>
> Rubin said the junior high school student was a "television addict" who often sneaked from his bedroom to watch all-night movies.
>
> Rubin said he intends to call as witnesses at the September 26 trial experts who will testify about "the impact of television upon the child, juvenile, and adult mind."

Television news reporters picked up the story as the trial progressed. The resourceful defense lawyer did call in his experts. The boy's mother testified that her son did indeed watch television during almost all of his free time. Here indeed was a candidate for a TV Addiction Clinic. And, if he had been acquitted, some resourceful psychologist would probably have started one. However, the jury didn't buy the unusual defense that television was the guilty party in the robbery-murder. Even if the boy had been exposed to countless crime shows, he is still accountable for his actions, the jury in effect proclaimed.

Many critics would disagree with the jury. Marie Winn in her book, *The Plug-In Drug: Television, Children, and the Family,* makes a scathing attack on *all* television. If the readers believe everything she writes in her book and magazine articles, America will one day witness the following scene:

> Early one morning the trashmen of America are startled to find a TV set in every refuse can. Some have their screens smashed in or their rabbit ears twisted off. Broken antennae are stuffed into many of the trash cans along with torn copies of *TV Guide.* The amazing story will be reported by Walter Cronkite that evening on CBS, but nobody will be watching. The whole nation has gone cold turkey in order to kick the TV habit.

Mrs. Winn does raise many valid issues which parents should ponder, preferably with the television set turned off. Like so many crusaders, however, she overstates her case, which leads to overdrawn conclusions. She gathers together TV "horror stories," reports, and surveys, and draws the worst possible conclusions from them. Her charges in many cases are true—that televiewing is often like drug addiction and leads to passivity, reduction of creativity, disruption of family communication, the centering of family routine around the TV set, and a host of other problems. To read her book is like hearing again the prohibitionists' tirades against Demon Rum: all the world's problems can be traced to its use. Prohibit it, and the problems of crime and poverty, divorce, child abuse, and all violence will disappear.

Neither naive neglect nor the passionate dumping of the set is a very realistic response to television. Television is such a fact of modern life that to reject it completely is to live in a cultural ghetto, to become virtually an exile in your own country. But to use it with no thought as to its consequences is to submit ourselves and our children to powers that will use us for their own ends—namely to make a profit from us regardless of what happens to us.

In our own family we have tried, often by trial and error, to follow a middle course. Television has become, in a sense, a member of our family, cherished and valued. But it's a dangerous member and has to be "watched." It's much like welcoming a tiger cub into the family. The cub is cute and cuddlesome. It's fun, and caring for it can teach us much about life and responsibility. But there comes a point when we had better watch out, lest it devour us. The contention of this book is that most American families have passed that point. Television has devoured them — their time, talents, values, and their possibilities of developing deeper family relationships. They are willing victims of a one-eyed member of the family whose dominance of family life is even symbolized by its place in most homes — the family room. It is also found in the bedroom, the kitchen, the dining room, the basement, the mobile camper, and even the bathroom!

Fortunately, the situation is not irreversible. Television sets require human hands activated by human thought and will to be turned on. If we can turn them on, then we can also turn them off. The watchword is the Greek dictum: "Moderation in all things." Television in moderate doses can add much to the enjoyment and enrichment of our lives. So can family games, reading, work projects, trips, and conversation. A balance of all these is necessary for the health of our family life, as a balance of vitamins is needed for a healthy body.

Parents set the tone for family living and establish the rules, especially when the children are young. Rules are set for brushing teeth, washing bodies, making beds, cleaning rooms, eating meals and a host of other activities necessary for family life. The one area too often not regulated is television. Few are the families where sharp distinctions are made between what is and what isn't "too much" television or between "good" and "bad" programs. Parents and children, too, usually have some idea of this in their heads, but fail to really talk it out and establish family norms. Thus judgments and decisions are made on the spur of the moment. This can give rise to arguments and resentments: "Aw, come on, Mom, I haven't been watching *that* long. Just one more program — please? All the kids will be talking about this one tomorrow!"

"*All* the kids!" How many parents have been trapped into changing their mind by this argument! It's one of the most potent weapons in a child's arsenal of defense against what he or she regards as unfair parental demands. Many are the parents who give in at this point, not wanting their child to be left out.

Discussion between parents (if there are two in the home) and then among all members of the family can avoid such conflict. A definite limit on the amount of time spent in front of the TV should be set, for parents *and* for children. This need not be inflexible, as exceptions will arise. It might be "ratings week" when the networks cram their schedules with all kinds of high quality programming, or a teacher might require that the students watch a TV show as part of their homework. But the reasons for making exceptions should be good ones, discussed and agreed upon by all. If parents say "No" to the TV set for themselves, children will learn to follow suit. If you were surprised by the results of your TV survey, the amount of viewing time should be thought over and discussed by your family.

The family should also discuss their criteria for determining which shows are good or acceptable. The discussion might be minimal when the children are quite young, but as they mature and ask questions, reasons should be given. And if they have reasons beyond the "all the kids are watching" one, you should listen to and discuss them. It just might be that there are good reasons for you to change your mind. Parents who turned off "All in the Family" because of "all that loud shouting and fussing" were often surprised at the depth and sensitivity of the writing and acting, when, as the result of a teenager's arguments and challenge, they sat down and watched one of the episodes.

When our children were small, we either switched channels or turned off the set when programs with too much violence or sex came on. We answered the pleas of the children with "We're sorry, we just don't think this is a good program to watch. There are many other things we can do that are so much better." When "Charlie's Angels" started on the air, our children were much older, so we told them our reasons for not tuning in — that it exploited women as sex objects, that the plots were totally unreal and offered no worthwhile treatment of issues or human relationships. In short, that it was phony and a total waste of time. There was a brief period of "forbidden fruit" feeling, since their friends were watching it. But then, their friends did many other things of which we did not approve — and their family life-styles were considerably more barren than ours, which the children could easily see. So they accepted our strictures. Recently, our two youngest, now adolescents, asked to watch an episode of the show.

Not worried now about their being conned by the series' values, my wife consented. "What did you think of it?" I asked upon arriving home and learning of this. "It really is stupid," one of the boys replied. "What a total waste of time!" There have been no further requests for this particular show.

No book can, of course, establish your criteria for viewing, but a few questions which you could ask about each program during a family discussion are:

1. What is this program really about?
2. What are the characters like? How do they relate to each other? Do they use people for their own ends? Are they stereotypes? Do they easily judge others as good guys or bad guys?
3. What methods are suggested as the right ways to solve problems?
4. Do the situations, even in a comedy, seem real? If it's a comedy, does the humor build people up or tear them down?
5. Does the program shed helpful light on an important social concern? Or does it exploit a current issue?
6. Will I be a better person for watching this program? Will I want to remember it a year from now?

We'll propose other critical questions in the following chapters. But the important thing is to begin to think about such matters and concerns.

Television is too important, and plays too dominant a role in our families, for us to approach it haphazardly.

CHAPTER THREE
Role Models on Television

If art is the imitation of life, then children should be good artists. Children learn by imitating those around them. Language and speech patterns, values and behavior, what is expected of them — all this, and more, children learn from others. Parents are the primary teachers but whoever comes into close and continual contact with children contributes.

Television is a powerful force in presenting models for children to imitate. TV offers role models of what it is to be a man or a woman in our society. These may or may not coincide with your ideals, but they are present and working on the mind of your child.

Researchers tell us that 4 hours a day is a low figure for most children's televiewing: that's 28 hours a week or 1,456 hours a year. During this time Mom is able to get chores done or rest for a few minutes knowing that her child is being cared for by the electronic baby-sitter—only it is a case of the baby watching the "sitter."

By the time the average child enters first grade, he or she has watched more than *five thousand* hours of television and has observed the actions and words of hundreds of TV people. What has the child seen and learned about what it means to be a man or woman?

According to most TV presentations, a man is to take charge, to be young, handsome and sexy, to be strong and tough and ready to fight. A man keeps his cool, is never caught off guard, or at least not for long. A man has all the answers for any problem. He comes out on top every time.

Think of the TV heroes that have been popular over the years. Matt Dillon and Kojak are two that are typical. Matt is tall and good looking, quiet, a man of few words and few outside interests. He wasn't overly aggressive, but when pushed he was quick with his fists or gun, whichever the occasion demanded. He could be depended upon. He would risk everything to help a friend. But as the very name of the show suggested, in the end violence was the only solution for most problems. Millions of Americans love Matt Dillon. So many that for years "Gunsmoke" was at or near the top of the ratings. It enshrined the mythology of the Old West and its image of manhood.

"Kojak," and most other such cop shows, is a western in modern, urban guise. Now it's the city that is the wild West. "It's a jungle out there," where only the law of the gun is understood. Kojak is given an endearing trait in that he likes candy and kids. But he also likes his fists and his gun. Violence is again seen as the best answer in fighting crime.

Kojak roughs up his suspects at times, and frequently sneers at them or abuses them in verbal ways. This is presented as acceptable because it shows that he cares passionately for justice, for law and order. A good lawyer would frequently be able to either tie up the case in court or have it thrown out on the basis of Kojak's violation of the suspect's constitutional rights. But such matters don't enter into our adult fairy tale. Kojak brings us comfort and hope. At least *his* portion of the city streets is relatively safe. He always gets his man (or woman). He's a *man*.

Matt Dillon and Kojak have been copied many times in countless TV adventure shows. For the very young, there's "Popeye the Sailor Man." The male formula has been very simplified in his case. No subtlety here, just an endless round of Popeye beating up Bluto who threatens Olive Oyle. Popeye needs the special help of his spinach (a not so subtle pitch for the favor of mothers?), but once he regains his tremendous strength from consuming it, he's a regular attack machine. Wham! Bam! Sock! Crunch! And our villain is appropriately destroyed. That's how a *man* settles his problems or disputes. And Popeye lets his girl friend know who's in charge, too. A man commands!

Our children watch and watch, and heed,

unless parents or teachers counteract the TV's message. Hopefully, kids see their father presenting a different role model. But this is often not the case, since he was raised with similar heroes— Hopalong Cassidy, the Lone Ranger, John Wayne, Dick Tracy, etc. It's a cycle that threatens to be endlessly repeated, as the tough, rough male is seen in comics, films, and books as well as on the small screen.

The cartoon hero successfully plays on children's fantasies. They wish they could stand up against problems and/or adults and tell them off. They wish they could have superhuman powers to smash their enemies. So they tune in Tarzan, Superman, and all the rest. There they see evil vanquished. The blood and gore of adult shows aren't as prevalent, but the moans and bodily violence are.

TV also presents role models for girls. Traditionally, this has meant being rescued by the male. Think of the endless westerns, police shows, and the kids' cartoons. How often do they portray a woman as a strong, able person? She is either weak, or she's dumb, in need of a *man* to protect her in this violent, evil world. In TV ads she is always being saved by the likes of Mister Clean, Big Wally, The White Knight and other male savior figures. Who gets blamed for "ring around the collar?" The poor, helpless woman, of course. In some commercials the male dominance doctrine is implied by the choice of announcers. The off-camera voice that supplies the solution to the poor woman's problem is always a *male's*. In a strong, know-it-all voice he offers the product that is "the answer."

Submissiveness is another trait in the TV woman. This is especially evident in TV commercials. Who is portrayed as serving whom those cups of coffee, delicious rolls, bowls of soup, steaks, etc.?

These TV stereotypes of male and female are repeatedly put before our children. What's a parent to do? It is important that parents present in their own lives a different role model. Men who can show their emotions, help around the house, and talk over problems will offset the TV stereotypes, especially if televiewing for the children is limited. Women who don't fall prey to all the advertising gimmicks, who work at jobs or for community causes, who use reason and do not give way to every fear will show sons and daughters another possible pattern of behavior for females.

By being aware of the TV stereotypes and pointing these out to children, the impact of the TV role model is reduced. Then, too, television does offer other models of what a man or woman is. These can be sought out and children encouraged to consider them. Wonder Woman may not morally be much of an improvement over Popeye, but at least *she's* a she—capable, concerned and coping—a far cry from Olive Oyle. Spider Man is still a superhero, reliant upon super powers, but he does resort to bodily violence less often than others. For the very young, there's even Casper, The Friendly Ghost. He'd rather make friends than scare people. He destroys his enemies by trying to make friends with them. Not a bad parallel to the ethics of Christ, when you think about it — and worth pointing out to children, too.

Adolescents have loved the Fonz of "Happy Days." His gestures, language, and dress have frequently been imitated. And his visage stares out from wall posters in many a young person's room. Although he's very different from the leather-jacketed hoods I remember from the real high school and college days of the fifties, he offers an intriguing role model for boys. He's cool and tough, which they want to be. But he's also the opposite. People who are *really* cool and tough aren't vulnerable. He cares about people. Despite his efforts to hide this, he often shows it. He's been known to cry for someone. He's not so tough that he doesn't realize that he needs help. When Richie was injured in a motorcycle accident, the Fonz prayed. He also isn't afraid to speak out for what he knows is right. When the others were knuckling under to parental prejudice by not showing up at a party where a black teenager was part of the band, the Fonz rushed in. His method of threatening with bodily injury all who didn't show up might not have been the best tactic, but his heart was in the right place.

For older teens, Lou Grant offers an interesting male figure. He's not young, and he's not a handsome Greek god. He is dedicated and sensitive, most of the time, to the feelings of others. He cares, not about looks, but about people. And intriguingly, he takes orders not only from the editor in chief, a male, but from the publisher, a woman. He often disagrees, but we see him talking and arguing. He shows how a man wins by marshaling his facts and reasons, and presenting them in a straightforward way. Often Lou wins, but not always. Life is that way. It's refreshing to find such a man on a TV series. He even makes mistakes at times and has to apologize.

"All in the Family" has presented an alternative role model of what it means to be a woman. Gloria is the modern woman, young and attractive, smart, subject at times to emotional fits, but able to cope. Even when she marries, she isn't about to submit without question to her mate. Edith Bunker at first was the traditional submissive wife. She was always at the beck and call of Archie: preparing the meals on time just the way he wanted them, getting his beers, and tending solicitously to his every whim. But as the series progressed, so did Edith. She was innately honest and concerned about others. When these traits clashed with Archie's, they emerged on top. It was often Edith who corrected Archie's tendency to larceny, whether in insurance claims, income tax preparation, or in keeping the money accidentally sent to them by a prune company. When Archie was going to call a plumber to demand that he not come back with his helper, because the latter was a prisoner in a work-release program, Edith forcefully intervened. There was an audible gasp, then a cheer from the audience, when she yelled, "Put down the phone, Archie, PUT DOWN THE PHONE, DAMN IT!" Needless to say, Archie complied to this unexpectedly strong demand.

Edith Bunker shows that a woman does not have to be young and beautiful to be an attractive person. In scene after scene, with Mike, Gloria, and Archie, Edith shows that she understands the human heart. She is the one who gives comfort and even halting advice. She is the one who effects reconciliation. She reaches out to neighbors, even those that Archie would reject. Edith Bunker is everything that the TV woman of most shows and ads is not, and we love her for it. In fact, her TV "death" in 1980 attracted the kind of attention usually shown to real life celebrities.

The TV male and female stereotyped role models can be dealt with if parents and others who work with children are aware of them, care about their effects on young growing minds, and do something about them. In a world of information overload and conflicting values, children need all the concerned help they can get.

Thus far we've talked about role models as presented through characters played in television shows. Television also presents many opportunities for children to meet real people. Fred Rogers on PBS offers his excellent "Old Friends . . . New Friends" which examines the lives of famous actors, writers, sports figures, and other notables — and often their families as well. Fred himself, as the soft-spoken host of "Mister Rogers' Neighborhood" shows that a man can be gentle and quiet, concerned about the feelings of others. He effectively counters the "macho myth" of masculine identity as portrayed on so many TV shows.

The talk shows, often on in the afternoons

when children come home from school, present famous people who have achieved something significant. Although these personalities are often on merely to advertise their latest film or book, the interviewer usually delves into their beliefs and background. These personalities are set before our young people as examples of what it means to be a successful man or woman. It is important that parents discuss this question with their children, so that they can have a framework within which to evaluate the life-styles they see on TV.

Shows such as "The 700 Club" and "The PTL Club" also present personalities that young people may emulate. These shows are carried by some UHF stations, by the growing number of Christian broadcast stations, and by channels of cable TV companies. (Put all these together, and you have in effect a fourth network—more on this later.) Patterned after the secular shows, the Christian talk shows bring to our screens a very different type of personality. He or she is usually a "star" —entertainer, sports figure, politician, or author —but with a very different message. Usually, a testimonial is offered as to how the person's life was changed by "finding the Lord." Such shows present God as a very real power, shaping and changing human lives; and yet critical viewing is important.

Labeling oneself a Christian doesn't mean that one's conversion is complete. Are some of these "Christian stars" merely trying to hype their product, book, record, concert series? Do they promise easy answers to deep and complex problems? Do they suggest that by following their example one will gain quick wealth and "happiness"—pictured, of course, as "blessings from the Lord?" It seems that many of these personalities haven't gone beyond their initial Christian experience and thus haven't seen that even Christians have to wrestle with deep issues and complex problems, even when they have the Lord's help. These shows do present many genuine, sincere Christians who have much to offer us. But they also include a few "wolves in sheep's clothing" that have always plagued the Body of Christ.

Young people need to be shown that Christians must move beyond simplistic slogans to deal with the pain and heartache of day-to-day living. They need to come to know Christians of deep, realistic faith, who are not immune to failure and suffering but who nonetheless persevere and emerge stronger than ever. Although too often it doesn't, Christian television can present such models sorely needed by our children in an age when so many celebrities seem to be interested more in the almighty dollar than almighty God.

CHAPTER FOUR
TV and Your Family's Values

Every parent and adult concerned with children should keep in mind that television teaches values. Every program and commercial presents the viewer with a set of values. We may think a show is "just entertainment," but there always is a set of values, and understanding of reality rooted in certain premises of right and wrong, underlying the program. The impact of these values might be more lasting on a young mind simply because they are contained in an entertainment show. Our guard is down when we're relaxing. But the way TV characters relate to each other, what they say and cherish, how they work out their problems—all stem from what the writers, producers, and directors believe about the world.

A classic example of this can be seen in the old "I Love Lucy" series that still is constantly being rerun on stations around the country. This once top-rated show is indeed funny. There's no questioning the comic genius of the writers and rubber-faced Lucille Ball. We laughed and laughed at her antics, as well as those of her bumbling neighbors, Fred and Ethyl Mertz. Hundreds of thousands of people, mostly children during the after-school hours still do. But many of the episodes have a questionable moral basis. Because everything always turns out well in the "I Love Lucy" episodes, the methods of Lucy are presented as OK — a classic example of the end justifying the means.

To say that a show is "just entertainment" is to deny that the show is teaching, is impacting the young mind with a set of values. No program is neutral with regard to values. All teach. The question is whether you are always aware of what is being taught in your home. And do you agree with the values of a particular show? Some of these will conflict with Christian values. Others, often surprisingly so, affirm and support the values we hold as Christians.

Let's consider some values unacceptable to the Christian that are being presented on TV. We must be aware of and counteract these if we, rather than TV, are to be the primary influence in our children's growth and development.

Deceit is OK

Many would probably rate acceptance of violence as the number one negative value presented on TV. But I think the approval of deceit by so many writers and producers is far more serious.

Deceit eats at our moral fiber, sowing mistrust and cynicism in its wake, and its acceptance paves the way for the decay of other values.

The "I Love Lucy" show is a prime example of how the practice of deceit is presented with approval on TV. The show was popular at the time of the quiz show scandals in the 50s. Both taught that it's OK to lie if you can get away with it. Many of the Lucy plots dealt with Lucy trying to get her way with her husband or trying to cover up some problem. She wants to sing at Ricky's night club. She's overspent her household allowance. She's bought something that she knows she shouldn't have purchased, and she's afraid of Ricky's scolding if he finds out. The writers were kept busy dreaming up plots and twists of crazy situations, calling for her to do funny, outrageous things. To hide in closets and boxes, to cover up mistakes by outlandish stories and explanations, to sneak behind doors and drapes and crawl under beds and furniture—all to prevent her husband from finding out the truth. Such antics are funny—but they're also serious in suggesting that deceit is OK if "no one gets hurt in the end." And, of course, in the 22 minutes of a TV situation comedy, everything turns out fine. All is made up for or forgiven in the last two or three minutes. Thus the end justifies the means no matter how deceptive they might be. Millions of children grew up thinking that the practice of deceit was OK, for their parents sat there laughing and gave no sign of disapproval. And this same plot of domestic deception was repeated in most other situation comedies of the time.

"I Love Lucy" isn't the only series that presents deception as acceptable. An incident in the popular family series "Eight Is Enough," suggests that deception is preferable to honesty among family members. Two of the grown daughters seeking to go it alone, move into their own apartment. This proves to be more expensive than they had realized. They want to return, but they can't admit their problem to their father. Dad knows of their plight, but he wants to protect their pride, so he puts up with his daughters' elaborate ruse involving the manipulation of their grandmother who had moved into their old room. The scheme "works," and the girls return home. Never is there an honest sharing between father and daughters, and thus no facing up to their situation, no possibility for growth in character and communication.

Evil is something "out there"

It exists in someone else, never within yourself. Police shows, westerns, and most cartoons divide the world into good guys and bad guys. Such a world view leads to very little understanding of ourselves or others. Indeed, it leaves us open to manipulation by those who profit from appealing to fears in a we-they, good guy-bad guy, world.

Children's programs especially perpetuate this view. Villains are evil, heroes are good. The hero is as good as the villains are evil. He might have super powers, but he is never tempted to misuse them. (Spiderman is the one exception to this that I have seen.) His heart is always in the right place, on the side of law and order. There is no suggestion of the complexity of human nature, of subtle temptations to evil that affect us all. The hero can do no wrong. All means that the hero employs are right because they are used to combat—no, to *destroy*—evil. This leads to the third negative value.

Violence is OK

So many shows on television present violence as not only the best, but the *only* way to destroy evil. After all, anything the bad guys get is well deserved. The Clint Eastwood films are the extreme form of this teaching. *Dirty Harry,* shown on TV, presents us with such a subhuman villain that we can hardly wait for the final scene when police officer Harry catches up with him. The audience applauded and cheered in the theaters when Harry kicks and stomps him to death. In most TV cop shows, the end result is the same— a hail of bullets pumped into the villain's body; a long, often slow-motion plunge from a tower or building; a fiery crash; a fall into a machine, acid pit, or pond with crocodiles. Only seldom is there a hint that other means might be more effective or humane. This is something worth drawing children's attention to by questions.

TV shows almost always overlook the corrupting influence of violence on the user. Jesus understood this aspect of violence when he said to Peter, "Put up your sword, for those who live by the sword shall perish by the sword." George Orwell warned us in his fable *Animal Farm* that those who use devilish (or pigish) tactics in the end become devils (or pigs) themselves.

In our family, we have always made it a point to avoid shows that contain a large amount of violence. Our children just did not watch "Gunsmoke," nor do they watch "Kojak" or "Starsky and Hutch" or even "Popeye." This doesn't mean that we avoid all shows with violence. This would be impossible, unless we would junk the TV set. But we at least control the amount of violence to which our children are exposed.

When our children were young, we made it a practice to talk about the acts of violence which they saw, especially if committed by the "good guys," to see whether there was an alternative. Could the hero have followed Jesus' teaching by turning the other cheek, offering love in the face of hostility, or by making the enemy a friend? Could reason or negotiation have worked? Was either tried? In a world of terrorist acts and "non-negotiable demands," someone should be suggesting St. Paul's "better way." If not Christian parents, then who?

Fortunately, children are exposed to such shows as "Casper, The Friendly Ghost," who offers an alternative to violence in dealing with hostility. Casper tries to reason with villains, even make them his friends. Casper holds to the prophet's dream of a world in which the lion will lie down with the lamb. He often is misunderstood by others. He frequently gets hurt. Jesus said "Blessed are the peacemakers, for they shall be called children of God," but he also warned of the cross to be borne, of misunderstanding and persecution. But Casper persists and eventually "conquers" nonviolently. As children watch Casper, parents would do well to reinforce the teaching that reconciliation, though less spectacular, is a better solution than shooting and slugging.

Uncommitted sex is OK

This un-Christian value pervading so many programs says that sex between a man and a woman is an appetite which should be satisfied just as you satisfy your body's thirst or hunger. Viewers are urged in subtle and not so subtle ways to freely indulge their passions.

There are limits to what is allowed on TV in depicting sexual scenes, but these are far different from the strictures of the 50s when a married couple in a situation comedy couldn't be shown in the same bed. Bedroom scenes had to have twin beds! Today, producers know that sex can mean big profits at the box office or success in the ratings game, and they work overtime to create as many scenes as possible that trumpet the hedonistic philosophy of unlimited, uncommitted sex. This view of sex is more pervasive on television than violence, and parents need to work carefully with children to counter its influence. Children will pick up your signals of approval or disapproval without your having to preach a sermon. As they grow older you can find opportunities to talk about the relationship between men and women, about the Christian concept of marriage, and about sex as an expression of committed love. Television will provide you with many opportunities to discuss these topics. It's up to you to take advantage of them.

Winning is everything

This value so often expounded on television contends that success is what life is all about. No one should settle for being number two. Life is a contest with winners and losers. No one wants to be a loser, so try to win at all costs. This kind of world view knows only competitors, not neighbors. It's a world view fostered by TV sports. It can take what should be a fun-filled learning experience, a Little League baseball game, and turn it into a grueling, hellish battle with parents cursing umpires and screaming at boys who strike out or drop the ball. If you win, you're OK; if you lose, you're not. Winners get treated to ice cream or pizzas—losers to a stern admonition to do better.

Although TV sports are the primary teacher of the gospel of winning, game and quiz shows also foster this viewpoint. Such shows use flashing lights, exciting or suspenseful music, cheering audiences, attractive prizes, and a wheedling master of ceremonies to lure the contestant on. The contestant should be excited and totally convinced that winning the prizes is worth everything, even the sacrifice of one's own dignity. The contestant is a victim of his or her own greed, excitable temperament, and of the producers who deliberately look for these traits when they choose those who will be on the show.

The belief that winning is all that counts furthers the approval of violence as a means to an end. The violent resolution of a conflict declares that there can be no reconciliation, only a winner and a loser. To seek reconciliation suggests that there can be two or more winners, or if reconciliation is not achieved, that all are losers.

One of the best episodes of the family series "Eight Is Enough" showed Tom and his family engaged in a football game against another family. Both sides took the game so seriously that it degenerated into a grim battle of win at any cost. Tom, himself at first caught up in the passion of winning, came to realize this and called a halt to the game. He called everyone's attention to what was happening to them. With this fresh perspective, everyone was able to return to the game and play it with enjoyment and in a more relaxed, friendly spirit.

Charles Schultz has consistently countered the win at all costs philosophy in his famous cartoon stories, over twenty of which have been televised. Charlie Brown is life's perennial loser—scorned by friends, constantly tricked by Lucy, captain of a baseball team that has never won a game, abused and used even by his dog at times. Yet he persists, something which parents should point out to their children as they watch the episodes shown throughout the year. Charlie is on the mound, even when the score is 98 to 0; he's out there with his kite despite that kite-eating tree;

he never gives up trying to kick the football despite Lucy's taking the ball away; and he frequently reaches out to people despite the rejection of so many.

The "Charlie Brown Christmas Special" is well worth watching and discussing together as a family, for it suggests that losers can be winners, too. Charlie Brown is sent out to find a Christmas tree for the gang. He returns with a scraggly little tree that no one else wanted. Neither does the gang. They ridicule Charlie and the little tree. Later, however, they think about what they have done. Each adds a decoration to the tree. Before they know it, they have turned the once sad-looking tree into a shining symbol of Christmas. They are reconciled with Charlie and his choice, for they now recognize the potential that Charlie saw from the beginning. Like Christ, another "loser" in the world's eyes, Charlie can see potential for value and beauty where others see only worthlessness and ugliness. Jesus recognized and drew forth beautiful qualities from society's losers — prostitutes, cripples, tax collectors, a hated Roman, Samaritans, the poor, even a thief on a cross. Christ's cross and teachings remind us constantly that we all can be winners. Only by trying to win power over others at all costs will we truly become losers.

Getting and owning things is what makes you OK

"You're nobody till somebody . . ." envies your new car or whiter-than-white wash. Television proclaims that life consists of getting and spending. The abundant life is measured by your bank account; the brand of your suits, dresses, or jeans; the number of cars, boats, or TV sets you own. Materialism is presented as *the* way of life. The abundant life is possible only with an abundance of things.

This is a long way from the life-style suggested by Jesus to the rich young ruler (Luke 18:18-23) or the rich fool (Luke 12:13-21). Again and again Jesus tells us that too much emphasis upon things will ensnare us, causing us to miss out on his kingdom. Occasionally, a television program itself will suggest this. Dr. Seuss's annual cartoon "The Day the Grinch Stole Christmas" shows the citizens of Whosville joyfully celebrating Christmas despite the Grinch's dastardly deed: he had stolen all their Christmas presents. The Grinch is puzzled and upset, too, since he was fiendishly looking forward to hearing the Whos wail over their stolen gifts. The villain's puzzlement gives way to the insight, "Maybe Christmas doesn't come from a store; maybe Christmas is something more." This program is positively subversive of commercial television's value system. Parents should watch this annual show with their children and talk about its message.

On public television Fred Rogers regularly stresses the value of people over things. Through song and story Mr. Rogers teaches that people are more important than things. His gentle efforts, reinforced by discerning parents, can do much to offset the rank materialism of most television.

Youth is the only stage of life worth living

You're over the hill if you're past 30, and if you are, you better not look like it.

Television is the land of the young. Only recently has the medium begun to deal realistically with old age in a few specials and dramas. Only one series has consistently presented an elderly couple in a dignified, enobling way—the grandparents in "The Waltons." (Probably no story was as touching and uplifting in dealing with the problems of the elderly as the episode of Grandma's return home after a stroke had sent her to the hospital.)

Cereal ads show a mother being mistakenly grabbed by the boyfriend of her daughter, so youthful is she as a result of eating the cereal. All of those who are shown using Geritol or Serutan or Polygrip look remarkably young. No doubt the result of Clairol or Grecian Formula and all the other magical goodies the commercials urge us to try. The fountain of youth has been discovered in video land; woe to those actually looking and acting their age!

Television is beamed at the young, with the old free to look on but not to get in the way. This state of affairs flows from TV's profit orientation. Network television is concerned with a mass audience. That's where the big money is. But not just any mass audience. Studies show that young people are the ones who spend most of the money in our society. Older viewers are more cautious in spending their money.

While television does promote the above-listed negative values, it does, fortunately, also affirm positive values. The values of caring and of family relationships are celebrated in many popular shows. "The Partridge Family" is still watched on reruns by many young teens and children because of the sense of family among the show's characters. A friend who leads youth conferences asked young people why they liked this rerun. "Because everyone in the family really cares about each other" was the reply. In real life these young people had seen very little of the love and support which they witnessed each night for a half hour on the tube.

"The Waltons" also offer a family with a strong sense of tradition and deep affection for each other. We see a family that helps each other through hardships and shares together in each other's joys. Certainly, the stories and characterizations on "The Waltons" are better than the overly simplified ones of another popular series, "Little House on the Prairie."

Thanks to producers such as Norman Lear and Grant Tinker, television has its own way of promoting the values of brotherhood and racial tolerance. Many of their shows have dealt with issues of prejudice, crime and violence, honesty and cheating, the problems of the aging, sexism and feminism, communication between husband and wife and parent and grown children, even of theology and the meaning of faith.

"M.A.S.H." holds forth the possibility that

even in the midst of war and regimentation, a concern for human dignity can be kept alive. The show's somewhat flippant thumbing of its nose at authority upsets some people, but this only highlights the importance of the human over the institutional, a value which Jesus also taught.

Such television shows can be seen as video parables, stories which point beyond themselves to a moral or spiritual truth. Norman Lear is a moral prophet reaching an audience of over fifty million. Most people don't realize that he is preaching because they regard Lear as "just an entertainer." Lear doesn't issue an altar call at the end of his programs, but he frequently is trying to get us to accept his viewpoint that what's inside people is important, not the labels which society uses to designate groups as "black," "women," "Mexican," etc.

Adults who nurture or teach young people would do well to be on the lookout for programs that qualify as video parables. These programs will stress the human over the institutional or legal. The characters will try to be open and honest with one another. They will not be paragons of perfection. They will have flaws that trip them up sometimes, but they will oppose efforts to dehumanize others. They will stand against injustice but not use unjust means themselves. These video parables will offer suggestions for facing the issues of our time, but with the understanding that there are no easy solutions.

You will not always agree with what is presented on television. This is to be expected. By not tuning in certain shows, or by watching them with the family and then presenting your own viewpoint, you can insure that your values and standards will be understood by the younger members of the family. Watching TV can be an occasion for your family to discuss an issue. (Or if you decide not to watch a program, you can talk over why you decided not to.)

The values and ethics of our young are too important to leave to the influence of uncontrolled television viewing. By remembering that the TV set in your home is *always* a "hidden persuader," an electronic teacher, and by understanding the value system behind television, you can insure that children will not be victimized. It's a big job, but when you realize that children will spend more time watching television than they will in school and church combined, you can see that it's a necessary task, even an exciting one.

CHAPTER FIVE
Children's Programming

Television offers the best and the worst in programming for children. There are fine, intelligent shows carefully put together by artists who love children. And there are cheap sloppily written and produced series that are ground out by people interested only in money. Parents should be as concerned about what their children watch on TV as they are about their children's diet.

Prime time for children is Saturday morning. All three commercial networks broadcast nothing but children's fare from early morning to early afternoon. Parents are supposed to be sleeping or using all those tools and supplies they saw advertised on TV to fix up, clean up things around the house. Television keeps the kids happy and out of the way.

It would be worth your time to sit down some Saturday morning with your children and watch the TV. They'll be surprised (mine were, anyway) and delighted if you don't come on too strong with heavy criticism. You'll probably see the following:

- A galaxy of superhero stars—Spider Man, Wonder Woman, Godzilla, Tarzan, Popeye, Plastic Man, Mighty Mouse. The characters come and go; the names change, but the content is pretty much the same. Some evil person or power tries to take over the world. Everyone, including the Army, Navy, Air Force, Marines, and the Police are powerless to stop the invincible villain. Only the superhero can save the day.
- Whacky weirdos. These are crazy creatures who constantly go through a series of high-speed misadventures. Bugs Bunny, the Road Runner, and Heckle and Jeckle are examples of these television counterparts of the old Keystone Cops slapstick type of humor. Lots of fluff and escapist humor here.
- Goofy and dubious heroes. Fred Flintstone, the Jetsons, Scooby Doo and Friends are kiddy versions of situation comedies and mystery-adventure shows in which the main characters are laughable. In a sense, they're often anti-heroes. They poke fun at the serious ones, and they help young viewers feel a little superior to the dumb, goofy characters they are watching. The scripts are filled with puns which are missed by younger viewers but fun for older children and adults.
- Public Service material. Bowing to pressure

from various groups, the networks offer a variety of short subjects throughout each hour. There are consumer tips on buying, even on watching TV commercials, brief news stories, and educational lessons.

- "School House Rock," for example, which deals with short lessons in science, math, grammar, and history. ABC claims that more children watch the 3 five-minute segments which they run on Saturdays, than all other educational TV combined. NBC produced a fine series, "Hot Hero Sandwich," which included music, interviews, and stories of famous people and how they coped with problems as teenagers, but unfortunately, they aired it during the noon hour, a very poor time for youth or parents on Saturdays.

Of course, Saturdays aren't the only time that programs designed for children are offered. Local stations offer both new network programs and reruns of old series after school. Lately there have been successful attempts to present high quality, youth-oriented dramas dealing with adoption, teenage drinking, and such. These after-school specials represent TV at its finest in dealing sensitively with real issues. If a parent is at home, he or she should watch such shows with the adolescents, as these offer valuable opportunities for a family to talk over problems. Such talks could help young family members be aware of potential problems in their own lives. When the working parent comes home, he or she can be brought into the conversation at the supper table. Children usually enjoy telling adults the plots of favorite shows, and the listening parent can then pose questions and offer comments to further the discussion.

Many evening programs are produced primarily for adolescents. "The Incredible Hulk" may seem ridiculous to adults, unless they recall their own fascination with *Dr. Jekyl and Mr. Hyde.* Young people are passing through a stage of incredible growth. They feel ugly and awkward at times. They reach for something, only to find that their arms are slightly longer, and so they knock it over. They feel intense emotions which they can't always control, hence the frequent clashes with parents. They retreat into sullen moods with a "Leave me alone!" They feel hassled, pursued by outside authority, and pressured by internal growth which they don't understand. Thus, the story of a man who has a monster inside himself that takes over his life whenever he is threatened is very appealing to adolescents. They can empathize with the hero, for not only is he threatened from within, but he is pursued by the police and a determined newspaper reporter who misunderstand his true nature.

This series, even when it goes out of production, will be around for many years on the rerun syndication circuit, so popular is it with young people. A parent who hasn't seen it would do well to sit down with the children and watch it. Opportunities for talking over fears and concerns can arise during the episodes. Many of the scripts are surprisingly well written, sensitively portraying the characters and their relationships. "The Incredible Hulk" is a far better series than its comic book origin might suggest.

As I write this, the kids' favorite cops and robbers show is "CHiPs." Girls are attracted because the stars are "cute." They're young, good looking, unmarried, and care about people. They are not as violence prone as are many TV police. They put up with a lot of provocations from the public before they swing into action. The action, of course, draws the boys. I haven't actually timed the length of the car and motorcycle chases, but it seems to be at least 25 percent of the show time. For adults watching, this seems like a great amount of repetition: shots of cars careening around corners, jumping over things, narrowly avoiding crashes. There are close-ups of wheels, motorcycle exhausts, faces with jaws set; aerial views of cars pursued by motorcycles; and scenes of spectacular crashes. People are seldom killed, just shaken up a bit before being hustled off to jail. "CHiPs" is fantasy, but it offers as role models fellows who like to help people. Our young people could do far worse.

Not many television programs attempt to arouse a sense of wonder, but "Star Trek" certainly does. In some cities, two or three stations show it in syndication at different hours. The crew of the starship Enterprise is a close-knit family who support and affirm each other. Captain Kirk is the patriarch who keeps things on an even keel. Spock adds a touch of mystery and by his rationality the assurance that there is a solution to every problem and an answer to most questions. Good science fiction appeals to our sense of wonder at what's out there beyond the planets and stars. It enlarges our vision by presenting us with the possibilities of other life forms, of different kinds of civilizations, of new vistas of reality such as space and time warps. "Star Trek" is, for the most part, good science fiction, a rarity on television. If your children enjoy the series (ours will watch an episode even when they've seen it three or four times), be glad and watch it with them. You might be surprised at how many episodes are like morality tales, filled with insights and provocative questions about life and its purpose. There's the faith claim that even though the Enterprise might press on to where no human has ever trod, it never outstrips the moral order of the universe.

"Mister Rogers' Neighborhood" is perhaps the finest series made for young children. Although his slow pace and easy-going manner of talking have been cruelly parodied by comedians and older youth, his style so thoroughly meets the needs of young viewers that parents should encourage the watching of this gentle series. There's none of the razzle-dazzle of the over-praised "Sesame Street" or "Electric Company." Working with a far smaller budget than the producers of these two hyperactive shows, Fred Rogers and company offer a time of quiet learning and fantasy that can lead to self-understanding on the part of his young audience. He's concerned not with just entertaining or presenting a jumble of facts, such

as letters of the alphabet, but with what goes on in the hearts and minds of the child on the other side of the televised image.

Working with psychologists (Fred himself has studied child psychology at the university level), he knows the mind of the young. Filled with a sense of wonder and questioning, the child is also subject to many fears. Everyone is so much bigger and more powerful. There are so many experiences that are fearful, such as the simple task of going to the toilet. The water rushes down so fast. What will happen if I fall in? And so Fred wrote and sang a song on one show assuring the child that he or she can't be flushed down the drain.

Silly to fear this? Maybe. Many adults laugh at a child's fears. Or they lie to a child and say that a visit to the doctor or the dentist won't hurt, when the needle or the drill might indeed inflict some momentary pain. Fred Rogers doesn't do this. He knows that fears are real for the child. He shows that he cares about the child, and through his brief talks, his little songs, and the delightful fantasy of his puppets in the "Neighborhood of Make-Believe," he helps the child to understand the fear and to discover ways to cope with it.

He really means it when he says, "I like you just the way you are." I've seen him in relationship with other children and talked with him about his work. My teen daughter went with me to visit him, and in the note of thanks *he* sent us, he asked about a school test that she had mentioned. He is as quiet and unassuming off camera as on.

Fred Rogers offers a laudable role model of

what a man can be. Children need to see that a man can be gentle and kind, sensitive to the feelings and needs of others. Adults would do well to study his approach to children. He does not talk down to them, nor does he discount their feelings. He often uses an indirect approach to a subject, similar to the method of Jesus' parables. He elicits responses by asking questions; he shows by example or in his puppet segments what he is talking about. A good example was the series on superheroes and monsters. He tells the children that TV monsters are just make-believe. He takes the audience to the set of "The Incredible Hulk" where we meet the actors, watch them put on their makeup, and learn about camera tricks. Then in the "Neighborhood of Make-Believe," the puppet story centers on fears and scaring people.

These are just a few of the programs worth watching, but what about the far more numerous ones that are just plain junk? What can be done about them? With very young children, this is best handled by turning off the set. "We don't believe that's a very good program." Reason, backed by firm determination not to give in, will eventually work with older children. By showing them why you disapprove of a series like "Charlie's Angels," for instance, you can take some of the force out of the protest. "This show says that women are just sexy and nothing else and that violence is OK. We don't think you should spend your time watching it."

This doesn't mean that you allow children to watch only programs that are "morally uplifting."

Some innocuous cartoons might be allowed in small doses, but definite limits should be set. Seek to make televiewing a quality experience, as often as possible.

Television can lead to reading. Some programs based on books suggest to young viewers that they obtain the book from their bookstore or library. A wise parent might check such programs in advance and "just happen" to have a copy of the book lying on the TV set when the show comes on. A good way to find out in advance about such programs is to subscribe to the following:

MASS MEDIA NEWSLETTER—Twenty-Third Publications, P.O. Box 180, Mystic, CT 06355. This twice-monthly publication reviews upcoming programs, as well as theatrical and educational films. There are frequently helpful articles on television.

TEACHERS GUIDE TO TELEVISION—699 Madison Ave., New York, NY 10021. Published twice a year, this fascinating journal reviews in depth major programs that can be tied in to a child's education. It carries articles on TV and children and often suggests books to read as follow-up to viewing a particular program.

The suggestion of turning the set off is frequently offered in this little volume. *Too* much TV, no matter how high the quality, is too much. We need to vary our activities, lest we become the dull, passive beings parodied in such films as "Being There." Watching television is basically a one-way affair with little opportunity for growth from interacting with other people.

When the television is off, parents need to provide opportunities for family activities. Picnics, games played at home or in the yard, visits to local attractions, making something together, cooking a special dish or dessert, biking or hiking, working on a hobby, or just plain sitting and talking—these are but a few of the possibilities for enjoying each other and growing in the process.

Although a certain amount of time is necessary for family interaction, the quality of time spent together is equally important. Fifteen minutes during which the parent is concentrating on listening to or talking with a child might be more rewarding than two hours in which an adult is present with a child but not really listening or interacting. Good families don't just happen. Families, like gardens, require planning and cultivation, if they are to produce a crop of caring children.

CHAPTER SIX
The TV "Advangelist"

Children watch television for entertainment. You would probably add information and education to your own list of reasons. These, however, are not the reasons why the network and station owners offer their programs. They see television as a business venture, a commercial enterprise. Sponsors think of programs as interruptions, necessary pauses, between their sales pitches. These sales pitches are most often couched in a "religious manner" to motivate us to buy. Today's evangelists promising instant salvation (or at the most "just a few seconds away") are Mrs. Olson, Cora, Mr. Clean, the good-looking white-jacketed men armed with "independent laboratory reports," and an entire tent full of others.

Whatever your problem, the TV "advangelists" have an instant answer, or rather, an "important message," designed for your backache, bad breath, unexpected company, dirty floor, dandruff, insipid tasting coffee, slippery dentures, a pet that won't touch "ordinary" pet food, obesity, whatever. Sin takes many forms, and many saviors are needed to overcome it.

Take a close look at some of these commercials next time and note how similar they are to the old Elmer Gantry style of evangelism. Sin holds you in bondage—let's say in the form of "bad breath." "If he kisses you once, will he . . . ?" ("When the roll is called up yonder, will I be there?") Do not despair! "Good tidings of great joy to all the people" are at hand. For unto you is given this day those round, flavorful discs which you can pop into your mouth. Just believe, that is, buy and chew, and that foul mouth is transformed into angelic sweetness. Brothers and sisters, we are delivered again from d-r-readful punishment and perdition!

Or look at the coffee ads featuring Mrs. Olson. The cardinal sin snidely pointed out to a quivering wife by a disgruntled husband is bad tasting coffee—certainly grounds for divorce. Poor wife, with her marriage and future happiness jeopardized, what is she to do? Enter Mrs. Olson, priest and minister, marriage counselor and evangelist all rolled into one. Good tidings she brings: salvation comes this time in the form of a red can that contains *the* coffee which overcomes human frailty, bad pots, mediocre water, and any other obstacle known to subservient womankind. She steps out in faith and pours the coffee. A few moments later hubby is pleased.

"Now *that's* coffee!" he benevolently croons. The believer is saved!

What about the ad in which the car-riding cowboy hero rescues the damsel tied to a railroad track. As he holds her in his manly arms, he declares, "Datsun sets you free!" To which she replies, "Datsun saves!" If *that* isn't religious language, then Billy Graham has never preached a sermon!

Such secular salvation is preached thousands of times everyday of the week. Whenever you turn on your set, you join the video congregation. If you think you can escape a sermon by staying home and watching a rerun or a football game, you're mistaken. These secular sermons are certainly shorter than those offered in church. Oftentimes, they're even more entertaining than the programs which they sponsor! But can they deliver on their promise of instant "salvation"? That, of course, only you the viewer can decide, but it's a question you should ask the next time an advangelist appears on your screen urging you to "buy some today." Consider the following passage from Isaiah:

> All you who are thirsty,
>   come to the water!
> You who have no money,
>   come, receive grain and eat;
> Come without paying and without cost,
>   drink wine and milk!
> Why spend your money for what is not bread;
> your wages for what fails to satisfy? (Is 55:1,2)

The TV pitchman is an advangelist. His company spends millions of dollars in research, film making, and purchasing air time to win converts. And it works. No new product has a chance of success unless the manufacturer invests in an expensive TV advertising campaign, usually reinforced by magazine and mail coupons and free samples. "As seen on TV" signs are taped to displays of food, toys, and gadgets in our grocery and department stores. But, as suggested, we aren't helpless if we understand what is actually taking place during those interruptions of our programs.

Adults can thus cope with the advangelist, but what about children? The very qualities that endeared them to Christ are the ones which make them vulnerable to the advangelist: openness and innocent trust, enthusiasm for the new, a love for color and excitement. The advangelist has learned this from the psychologists hired to study child behavior. Millions of parents unthinkingly deliver a willing audience to the advangelist every day of the week, and even more so on Saturday mornings.

The advangelist uses cartoons, catchy jingles and slogans, magnificent photography, and top actors to sell the secular gospel to the child: If you want to be really happy or in with the gang, if you want to be important or have fun, then you must buy *this* product. Happiness is found in things.

Jesus warned long ago of the dire consequences that will befall those who cause "one of these little ones" to stumble. Parents, godparents, and the Christian community have the responsibility to show the child a better way to happiness. We discovered that you can't begin too early. When our oldest child was two or three, the cigarette ads were still on. One day when he overheard us talking about our parents and their smoking, he informed us that smoking is not bad. We inquired why or how he knew this, and he replied that the man on TV told him so.

Thus began our own efforts to counter the advangelist's influence on our children. We've found that a healthy dose of skepticism is a must if children are not to be victimized by the advangelists. One day the pastor's three-year-old daughter visited us. She was constantly humming and singing. She was a child of the church, but it wasn't hymns that she was singing. She was singing about Lestoil and cigarettes and Pepsi Cola and Milwaukee beer and so on. She liked the catchy TV jingles. They were repetitious, they rhymed, and were easy to memorize. In fact, you almost could not forget them, even if you tried. The advangelists are smart and clever. If they can get them young, they think they can hold onto them for life.

And they will, unless adults are alerted and do something to offset the advangelist's message. Your first temptation might be to set the kids down and give them a good talking to about television and advertising. This is not likely to do much good. Better to talk with the child as you watch television together. Make a brief comment when the situation calls for it. Better still, raise a question or two to stimulate the child to do his or her own thinking and evaluating. This will be more natural and, in the long run, more effective.

Some questions/comments that might prove useful in talking about TV ads are:

- What is it in the ad that grabs attention? The music? The color? The cartoon style? The promise of excitement or pleasure? Louder volume? Humor?
- Do you use any of the products shown? Did you first learn about them on TV? If it's a toy ad, and you have the toy, does it still work? Do you still play with it? The TV ads show mechanical toys always working perfectly. Ask your child, "How do you think they got them to work so well?" You can explain how a scene can be shot many times and then only the best shots selected and spliced together, giving the illusion of perfection. Does your child become bored with walking dolls and such? Are Raggedy Ann and Andy ones just as much fun? Why? Are building blocks as much fun as the expensive electronic gadgets?
- Do you think food ads exaggerate the quality of snack foods? Are they as good for you as milk and fruit? Why does the advangelist try to make it appear as if mothers are always glad to serve up such snacks?
- What medicines do you see advertised? Do you think they work as well as the man in the white jacket claims? Is he a real doctor or an actor?

CHAPTER SEVEN
Coping with Television

We've looked at many of the effects of television on children and made suggestions for dealing with television as a member of the family. Television, we've warned, can take over a family, dominating its life and negatively influencing its values. The first step in coping with TV in the family is to care deeply about the quality of family life and to want to protect or enhance it. The second step is to become informed, first by discovering just how much television is involved in the life of your family, and then by studying and talking about the medium itself.

Regardless of how good the content of television is, too much TV, like anything else, can be too much. Action for Children's Television, A.C.T., offers a delightful way to be reminded of this. They will give, for the asking, a large red tag that reads: "REMINDER — Too Much Television Can Be Harmful to Your Child." With the tag is a set of suggestions on how you can help your child limit and still benefit from TV watching. The tag has a string with which you can tie it to the TV set.

A.C.T.'s very name suggests the third step necessary for caring adults to tame the TV set — Action. A.C.T. has many action materials and suggestions that can help adults deal with TV and children. A free chart is entitled "Treat TV with T.L.C." The famous acronym this time means *T*alk, *L*ook, and *C*hoose. This small poster when taped on the refrigerator or the family bulletin board can serve as a constant challenge to the family. A.C.T. suggests that families talk about the programs, the characters, the way they relate, the methods by which they resolve problems, and even the ads. Such discussions can help children become more discriminating viewers.

"Look at TV with your child," A.C.T. continues. Look for dangerous behavior the child might be tempted to imitate. An episode of "Eight Is Enough" showed the youngest member of the family mowing the lawn in his bare feet! I wonder if the writers or producers thought about all the children watching this who would be tempted to take off their shoes while working on the lawn. A cereal ad, wanting to show how "natural" its ingredients are, showed a famed naturalist walking through a field. He picked several plants and ate them. The ad had to be taken off the air when reports of children being poisoned began to come in. They had gone out and started eating plants, imitating the "TV man." Unfortunately, they lacked his knowledge of what is edible and what is harmful.

Look for positive examples your child might imitate. Which programs depict people caring about each other? Are women portrayed as strong and forceful, good at their work, and not just good looking? Look for people from different racial or ethnic groups. And then look beyond TV — look for alternatives to televiewing such as games, trips, and hobbies.

Last of all, the A.C.T. chart suggests that you "Choose TV programs with your child." Choose the kind of programs and the number of programs to watch. Choose to turn off the set. Choose that great alternative to network television, public television. Too few people are aware of the high caliber of programs offered by PBS. History, nature, geography, science — these and much more are offered by PBS in entertaining formats.

Choose to respond to television by writing the station, the network, or an advertiser. If you don't like something, write a carefully reasoned letter stating your opinion. Simply state your reasons for disapproving. Also, if you are especially pleased or inspired by a program, let the people responsible know. They receive very few such letters and will appreciate them and be encouraged to continue such programming. When NBC aired "Jesus of Nazareth," the finest portrayal of Christ ever offered on TV, I asked the people of the Pittsburgh station how many letters of appreciation they had received. "Not one," was their reply. No wonder television people sometimes wonder if anyone out there really cares about quality programming!

But does writing letters really make any difference? Virginia Carter, Norman Lear's assistant, told a group that at the height of the popularity of "All in the Family," they received less than two dozen letters during any given week, from an audience of over 55 million! She reported that copies were made of the thoughtful letters with specific criticisms or suggestions. These were circulated among the writers, directors, and other staff members for their consideration.

TV can be changed by such efforts, as A.C.T. itself can attest. Action for Children's Television has been able to have the networks reduce by forty percent the amount of commercials shown on weekends. They helped eliminate the advertising of vitamins on children's programs, thus preventing poisoning by vitamins that children had been taught were like candy. A.C.T. also conducts intensive research and often presents its findings to the Federal Communications Commission in an effort to influence television producers and advertisers.

By representing the concerns and efforts of thousands of parents and professionals, A.C.T. has shown what a united effort based on intelligent research and action can accomplish. You might want to consider joining yourself. You'll add your name and a little money to that of others. You'll receive a quarterly newsletter about developments in children's broadcasting and what A.C.T. is doing. You will also have access to A.C.T. publications presenting the findings of their research.

You can also join together at a local level to discuss TV and pool your efforts in dealing with it. Church study groups, PTAs and PTOs, book study clubs—all are natural forums for such study and action. You could give book reviews of this and other TV books. (*The Plug-In Drug* would be a good one to pair with this.) Such efforts could awaken those adults who still think of TV as "just entertainment," and who see no harm in letting their children watch TV for hours at a time.

The national PTA has been involved in a campaign to get parents concerned about children and television, and will readily assist in launching a program to deal with the issue. A local PTA might have a presentation on the subject and then send home weekly flyers with information on the impact of TV on children.

A good source for information on the educational uses of television is the Parent Participation TV Workshop Project of *Teachers Guide to Television*. Recognizing the great educational potential of many television programs, these people offer materials to parents and teachers willing to gather together to talk about television and then to send back a short report. Participants receive advance information about worthwhile programs and a packet with articles on television and discussion guides for specific programs. One packet included guides for "Brave New World," "A Woman Called

Moses," "Heidi," "Les Miserables," "The Gift of the Magi," "Happy Birthday, Charlie Brown," "The Corn Is Green," "Roots: The Next Generation," "Make Believe Marriage," "Backstairs at the White House," and "The Long Journey Back." These were all excellent programs for adults and children or youth to watch together, and the use of even a portion of the guides would have enriched the viewing experience.

These materials are free to those willing to organize a viewing and/or discussion group. The group could meet at a school or in a home. If you cannot interest the school or PTA authorities in such a project, then local church leaders might be open to your suggestion to initiate a Parent Participation TV Workshop Project. The resources section of this book lists sources for further help.

You can launch out on your own by gathering some friends and neighbors together for a series of discussions. Have them make a profile of their TV viewing habits. They might be very surprised at how much, and what kind of television they are watching. And if they have children, they can be challenged to become aware of what television is doing to their family life. This doesn't have to be a heavy discussion; it can be a time of fun. Here are some games you can play either in your own family or as a member of a TV viewing discussion group.

1. "Spot that Stereotype." Television programs and ads are replete with stereotypes. Stereotypes are shorthand methods of conveying ideas

about people. They can be amusing, or they can be harmful to the minds of children, especially if the stereotype incorporates racial, sexual, and ethnic myths. Stereotypes foster the illusion that people are simple machines rather than complex personalities, often possessing conflicting traits and emotions. Some stereotypes to look for:

The dumb/helpless woman — She inhabits most cartoons, ads, and male-oriented adventure shows.

Pure hero — He never has a doubt about his cause or his methods, always wins, is master of every situation, and reveals no weaknesses.

Heartless villain — Totally evil, possesses no redeeming traits, comes to a violent end. A subcategory is the evil foreigner — swarthy, treacherous, utterly ruthless.

The good/fallen woman — She's usually a prostitute but has a heart of gold. She fails to win the hero who is destined for happiness with the pure heroine.

The scientist — Depending on the producer's point of view, he may be the mad genius who threatens the universe, or he may be the savior of the world.

The coward — Most cartoon adventure shows have one of these scared-of-his-own-shadow characters.

There are many more stereotypes you can add to the list. You can play by seeing who can spot the first stereotype, which means interrupting the show. Or the group can watch an entire show, ads included, then turn off the set and see who has listed the most stereotypes. Each person should be prepared to describe the stereotype or to give reasons why he or she thinks the character is a stereotype. This could generate some good discussion, as not everyone will agree that a particular character is a stereotype. For example, are the characters of "All in the Family" stereotypes? Archie Bunker is often seen as the arch-stereotype of a bigot, yet many liberals have complained that he is often shown in too favorable a light. I think this is because he is often shown as the complex human being that he is—bound by his prejudices, yet possessing generous impulses to think and do good. Both he and Edith have grown through the years, as real people do.

2. "What's the Promise?" While watching a commercial the group tries to figure out what the advangelist is promising to those who use the product.

   An example: "If he kisses you once, will he kiss you again? Be certain with . . . " What is the advangelist appealing to? The dream most of us have of being attractive to a member of the opposite sex? No one wants something like "bad breath" to get in the way. So, the ad promises us success with the first kiss, as if the product alone will make this happen.

   Still another example is the coffee ad which implies that marital success depends on the wife's being able to make good coffee.

   Sometimes the promise will take keen discernment, but it will be there. The advangelist knows our dreams and our weaknesses, and this game can make a group or family aware of the techniques used in commercials to seduce us.

3. "Blessed are . . . " This is for those who enjoyed "What's the Promise?" and are willing to pursue the idea a bit further by connecting it with the Scriptures. Read the Beatitudes of Christ aloud (Mt 5:1-12) and discuss them in relationship to the values and promises ("Blessed are . . . ") of TV ads. What does Jesus promise? Are the qualities extolled in the Beatitudes those that the world usually thinks of as desirable? Will they lead to success in the world's sense? Spend a few minutes (perhaps with the group divided into twos or threes) and see if you can come up with some Beatitudes According to Madison Avenue. Some possible ones to get you started:

Blessed are the big spenders, for the whole world belongs to them.

Blessed are the VISA and Master Chargers, for they shall inherit the earth.

Blessed are those who gorge themselves with rich foods, for their stomachs shall be comforted with Alka Seltzer.

Blessed are those who hunger and thirst, for they shall be satisfied with Coke, Pepsi, or Bud.

Blessed are those who fear persecution from bosses or peers, for Right Guard or Ban shall protect them from all social rejection.

Rejoice and be glad at all the riches of the department and grocery stores, for your reward is great in the Kingdom of Video.

The above can actually be a good way to study Scripture and at the same time relate it to what is being expounded by television. Other portions of the Sermon on the Mount and parables, such as "The Rich Fool," could be discussed in relation to television.

4. "You Write It." In this game, you watch a program, preferably a half-hour comedy or adventure show but without the sound. The group makes up the dialogue, and even sound effects, as the program progresses. This can have everyone in stitches, and the laughter won't be canned.

5. "And in Conclusion . . . " Many situation comedies and adventure shows have contrived endings. If your family is watching such a program, turn off the set and discuss some possible endings. Talk especially about the methods used in resolving conflict: Are there better ways? Act out these alternate endings if you feel comfortable and enjoy doing so. Keep reality in mind. Do people change as rapidly as in the program? Is the answer to a problem really as simple or easy as depicted? Can evil or dishonest means produce good results?

* * * * *

Thus far in this chapter our focus has been on commercial television, but there are alternatives, namely, the Public Television System (PBS) and the large number of Christian programs, sometimes loosely lumped together as the "Fourth Network." Actually, PBS should probably be called the "Fourth Network," since it's been around a little longer. Starting out as "educational television," PBS now offers programs as exciting and of as high quality as those offered by major networks. Adaptations of classic novels and plays, musicals, classic films, interviews and news, snappy children's shows—they're all there waiting for a largely unaware public to discover them. PBS's audience is tiny compared to the other networks', but consider that more people see one of Shakespeare's plays during a PBS broadcast than ever saw a production during the Bard's entire lifetime. Its impact is important. If you haven't introduced PBS to your family, you should soon—even if you have to install a special antenna to bring in the nearest station.

In an earlier chapter, mention was made of Christian television. This is a growing, dynamic force that currently appeals mainly to evangelical Christians. Pat Robertson, head of CBN (Christian Broadcasting Network) has often been quoted as saying that he is attempting to set up a network to rival the commercial ones. With millions of dollars to work with, most of it received from the viewing audience, he is well on the way to achieving this. CBN boasts modern television production studios as good as commercial ones, trained, dedicated technicians and personnel, and a satellite system that can beam CBN programming to virtually any spot on the planet. When you add the efforts of evangelical ministers Oral Roberts, Robert Schuller, Jerry Falwell, and Rex Humbard, to name only the front-runners, you have quite an array of Christian media "stars." Some of their programs appear on secular stations, and almost all of them can be found on "Christian" radio and television stations and cable systems.

This movement to establish a "Christian alternative" to commercial television has drawn a great deal of attention, in and out of the churches. Some critics are worried about the effects of what they call the "electronic church" upon the local church. Media evangelists finance their programs, or ministries, by appealing for funds from the viewers. Will this siphon off money desperately needed by local churches? And is the competition with the local parish unfair since few of them can afford the large choirs, talented soloists and special speakers, the special lighting and all the other accoutrements of the spectacular TV worship service? Will people drop out of the local church as they tune in to the electronic church? After all, if you don't like the personality or the theology of your local priest or pastor, all you have to do is turn on your television set on Sunday morning. And you don't have to dress up and sit on a hard pew or run the risk of being asked to serve on a church committee.

Such concerns are very real and yet to be answered fully one way or the other. But for the purposes of this book, one additional word needs to be said. Even if your family watched only Christian television, the matter of the amount of televiewing must be considered. It could be almost as disastrous for a family to sit in front of a TV set for 5 or 6 hours a day viewing only religious programming, as to spend the same amount of time watching secular junk. The *amount* of viewing time should be controlled as well as the content.

Television is a fact of life — an important fact, but *TV is not life*. It need not control our lives, if we are aware of its influence and choose to do something about it. This little book has presented you with a few ideas and techniques for governing your family's life in relation to television.

Television can either blight or enhance family life. The choice is up to you. The best defense against the former is to develop a well-rounded family life, only a part of which is watching television. Families have changed over the past fifty years. The automobile, the Depression, wars, radio, movies, and television all contributed to the changes. New developments in television will contribute much to changes in the next ten years. Videodisc machines and videotape recorders will cut into the influence of commercial TV networks. A family need not be enslaved to a network schedule if they own a video recorder. The machine can record a desirable program while the family is out enjoying a picnic or a concert. With a camera, the price of which continues to decline, the family can produce its own TV "movies." The potentials of cable television are still being developed. Soon there will be widespread opportunity for two-way communication by means of cable systems, such as the QUBE system in Columbus, Ohio.

The new technology will bring the danger that with so many films and programs to watch and with new means for enjoying or participating in television, a family will spend even more time in front of the set. The statement that television should be just a part of a family's life needs to be repeated. Participating in active sports, trips, school and church functions is a necessity for family growth. Children need the love and attention of warm, lively parents—not a wired picture box. *You* and not television are the key to their future. It is your responsibility before God to commit yourself to nurturing their tender minds and spirits.

Suggested Resources

I. Books

Alley, Robert S. *Television: Ethics for Hire?* Nashville: Abingdon Press, 1977.

    A good profile of those primarily responsible for TV shows, the TV producers.

Logan, Ben, ed. *Television Awareness Training: The Viewers' Guide for Family and Community.* Nashville: Abingdon, 1979.

    This handbook for the Television Awareness Training Workshops contains helpful articles on TV and values. The communications office of your diocese or church regional headquarters probably has this and can recommend someone trained to lead a TAT workshop.

McNulty, Edward. *Television: A Guide for Christians*. Nashville: Abingdon Press, 1976. Now available from Visual Parables, First Presbyterian Church, South Portage, Westfield, NY 14787.

A resource guide for groups wanting to study TV, values, and Scripture. Included are many ideas for group viewing and discussing of soap operas, adventure and children's shows, sitcoms, ads, and TV news. There are even worship celebrations built around TV themes.

Marsh, Spencer. *God, Man, and Archie Bunker*. New York: Harper and Row, 1975.

_____. *Edith the Good*. New York: Harper & Row, 1977.

These two books by Spencer Marsh are excellent examples of the application of biblical/theological insights to a TV series. Both books contain photos and generous portions of scripts from various episodes of "All in the Family." You can obtain an audio cassette for the first book. In it Spencer Marsh leads a group through six sessions that explore basic Christian doctrines as illustrated by scenes from the TV show. The cassette which also includes excerpts from the sound track of the series is available from THESIS Theological Cassettes, P.O. Box 11724, Pittsburgh, PA 15228.

Norback, Craig T. & Peter G. *TV Guide Almanac*. Ballantine Books: New York, 1980.

A fascinating reference work that includes the names and addresses of the top 100 TV advertisers (and the amount they spent in 1978), the addresses of all TV stations, the Television Code of the NAB, a section on the FCC, a brief history of TV, and much more.

Winn, Marie. *The Plug-in Drug.* New York: Viking Press, 1977.

A spirited, often intemperate attack on television; it should nonetheless be in your library.

II. Periodicals

Cultural Information Service. P.O. Box 92, New York, NY 10016.

A biweekly publication surveying the cultural scene—television, novels, pop music, and film. CIS often provides fine discussion guides for specific TV specials and films.

Help for Television Viewers. The Christian Life Commission, Southern Baptist Convention, 460 James Robertson Parkway, Nashville, TN 37219.

This is actually a packet of materials designed to help the viewer watch TV more discriminately and take action to improve TV. The packet comes with discussion guides, TV survey form, and a listing of resources.

Mass Media Newsletter. P.O. Box 180, Mystic, CT 06355.

This biweekly publication of Twenty-Third Publications will keep you abreast of new developments in TV and inform you about upcoming TV programs, as well as films and other audio-visual materials. Written from a religious perspective, the articles and reviews will be very useful.

Teachers Guide to Television. 699 Madison Ave., New York, NY 10021.

Although published only twice a year, this journal is most helpful in relating childrens' televiewing to their school and classroom experiences.

III. Organizations

Action for Children's Television (ACT), 46 Austin St., Newtonville, MA 02160.

National Association for Better Broadcasting, P.O. Box 43640, Los Angeles, CA 90043.

National PTA TV Action Center, 700 North Rush St., Chicago, IL 60616.

Parent Participation TV Workshop, c/o *Teachers Guide for Television,* 699 Madison Ave., New York, NY 10021.

Television Information Office, 745 Fifth Ave., New York, NY 10022.

IV. Television Networks

ABC, 1330 Avenue of the Americas, New York, NY 10019.

CBS, 51 West 52nd St., New York, NY 10019.

NBC, 30 Rockefeller Plaza, New York, NY 10020.

PBS, 485 L'Enfant Plaza West, S.W., Washington, DC 20024.

V. Films

These films are a good means for initiating a study of television and its impact on viewers. Available from A.C.T., 46 Austin St., Newtonville, MA 02160:

It's as Easy as Selling Candy to a Baby — 11 min., $25 rental.

A TV announcer introduces scenes of children watching television as they eat candy bars. The TV ads promise fun and excitement with each bite, but a children's dentist shows us a child whose teeth have been ruined by too many sweets. The scenes present a strong visual warning to parents about the power of TV advertising over children.

Kids for Sale — 20 min., $30 rental.

This film shows the seductive power of Saturday morning TV ads, and features parents who give advice on how to guide young viewers to understand and withstand TV's persuasive powers.

Available from your local film library, media center, or Mass Media Ministries:

TV: The Anonymous Teacher — 15 min., ($20 from MMM).

A fascinating look at the effects of television on young viewers. We not only hear from experts, we see children's faces react as they actually watch TV. In one corner of the screen we see the TV program the children are watching. This is a very persuasive film showing children actually imitating the TV violence they have been viewing.

The 30-Second Dream — 15 min., ($25 from MMM)

This well-selected collection of top TV commercials demonstrates TV advertising's seductive powers. The narrator points out the techniques used to play on our hopes and fears so that we'll "buy some today." There's a printed guide available with the film that will help a group explore the values behind TV ads and the real or imagined needs they address. Excellent for junior high through adult ages.